Guidebook For Your Digital World

ACKNOWLEDGMENTS

I would like to thank the many people who helped this book become a reality. I would not have been able to get it done without the love and support of my wife and children, who gave me the encouragement and time to research and write this book.

My mother played a large role, not just as my target audience for the book, but also as my editor through the multiple versions and drafts needed to make this book a reality. Thank you for a lifetime of support.

Finally, I must acknowledge that my career has been guided along the way by several managers that believed in me and pushed me to do things that I didn't think I could do. Thank you for pushing me to become more than I thought possible.

(And I can't forget the sandwich that started my whole IT career...)

Guidebook For Your Digital World

How To Avoid the Scammers and Pickpockets on Your Digital Journey

Chris Dailey

Chris Dailey

CONTENTS

Introduction

My interest in security began with my own physical security. I wanted to know how to recognize what was going on around me so I could stay safe. I've always been given advice like "be aware of your surroundings" or "keep your head on a swivel" but I wanted to know what to look for when I was keeping an eye on my surroundings.

I began to study different types of thieves to see just how they operated. How could a pickpocket remove the glasses from someone's face without them knowing? I became fascinated with the techniques used by pickpockets and magicians to control the attention of their target.

One day this all came together in a two-second encounter with a pickpocket team in a subway under the streets of Rome. I recognized that I was being targeted. I knew the techniques that the team was using. I was able to pull someone's hand out of my pocket in time to save my vacation.

Over the years I worked towards becoming an expert on

recognizing and avoiding techniques used by an attacker. I started to teach the National Rifle Association's *Refuse to be a Victim* program to help others recognize how they could be targeted and build a plan to avoid looking like an easy target.

Once I started working in cyber security, I went looking for a similar program and a better way to teach people how to stay safe in their digital lives. I never found the program I was looking for, so I finally just sat down and started writing.

This is not a technical book on cyber security. If you're looking for help getting into *HTB* or studying for your *CISSP* (or even know what those acronyms mean), then this book is **NOT** for you. I wrote this book for my mom, and all the people like her that feel like a tourist in the digital world.

We are going to go on a tour through the digital world and point out the potential dangers. We are going to talk about how attackers look for victims and how you can stay off their radar. We may take side trips through technical terms, but I'll be by your side the whole time to explain what the term means and why it matters.

When we're done, I hope you feel comfortable in the digital world so you can feel like a local and enjoy the journey.

Passwords

Let's start our journey with a security concept that goes back at least as far as ancient Mesopotamia and probably beyond – the password.

We use them every day to verify that we are who we say we are. Some services are trying to move away from passwords, but I don't think they're going away any time soon, so let's talk about how an attacker will use your password against you – and how to stop them.

In the early days of the Internet we were a little naïve about how we stored passwords. Website owners would create a file with a list of users and their passwords that they could check when you logged into a website. Attackers eventually started to find the files and we had the first data leaks.

Software engineers eventually developed a way to create a fingerprint of a password and store the fingerprint using a process called hashing. (See "What is Hashing?" at the end of this chapter for a better understanding of how this process works.) These password fingerprints are one-way and cannot be decoded, so when you log into a website the website checks the fingerprint of the password you entered against the password fingerprint stored in the user file.

In a perfect world the password problem would have been solved, but attackers kept getting more creative. They started collecting old data leaks and putting together databases of users, passwords, and the hashed password fingerprints – opening new ways to attack users.

Password Stuffing

This is an attack that takes advantage of the fact that we all have our favorite passwords that we can remember and like to use, so that's the password that we use every time we sign up for a new service.

Attackers compile a list of usernames and passwords from leaked websites and use them to login to other sites. When a new service launches there's usually a story that comes out within the first week about how the site has already been hacked. What really happened is that one or more attackers have gone through their list and found accounts for users that have reused the same password.

The website was always secure, but those user accounts are not secure because they used a password that was leaked from another website. We'll talk more about selecting a password later in this chapter, but it is important to use different passwords for each account that you set up.

Rainbow Tables

Unfortunately, websites still leak data. We would like to think that the stored password fingerprints are secure, but computers can store and process more data every day.

Attackers now create files called "rainbow tables" that contain passwords from previous leaks and the fingerprints for those passwords. They will also create different versions of those passwords (i.e. "password", "passw0rd", "p@ssword", etc.) and store the password fingerprints for every variation they can think of. As data leaks happen, they will use their rainbow tables to look up the password fingerprint to see if it is a known password.

You can make your passwords difficult for an attacker by selecting a strong password and unique password for each site so that it will not be in an attacker's rainbow tables.

Selecting a Password

Security professionals have debated about what makes a good password for years, and different places have different password policies, but I'll give you my tips on coming up with a secure password.

A password should be:

- Longer than just a word – a pass phrase is less likely to be in a rainbow table than a single word.
- Unique – you should not use the same or similar passwords across sites.

There are different ways to meet these goals and you can pick the one that works best for you. I'm going to share some that I use and that I recommend to others.

The ideal is to create a random password for each account that you set up. There are random password generators that will create a long and random password you can use. There are also sites (like dinopass.com) you can use to create passwords that are random variations on common words or phrases that are easier to remember.

Of course this means that how you keep track of your passwords is important, but we'll cover that in a minute.

Some people are less comfortable with randomness, so I've found a method that works well for the randomness-adverse people that still need strong passwords. I keep a hymnal by my desk that I can use to generate passwords when I need one. I might flip to a page and decide that my next password is going to be "063.AbideWithMe". This will meet the complexity requirements for just about any website. It is long and unique and all I have to remember (or store somewhere) is that the password for the site is 63. You can do the same with books using the page number and first line, but hymnals are usually easier to find online if you forget the password.

Managing Your Passwords

If you are going to use a new password for each site then you are going to have to keep track of your passwords somewhere. There are a wide variety of ways to store your passwords to help you so let's talk about some of my favorites.

A password manager is a software application that will store your password securely online. They often offer features like family plans, password sharing, and storing other secure information in one place.

When you use a password manager it will store your accounts and make sure that the information is secure. When you create a new account it will create and store a random password for you so you don't have to remember it. When you return to a website it will fill in the login form with the password you saved. All you need to know is the one password that you set up to unlock your password manager. It should be a very secure password you can remember and one that you have never used on another site.

The drawback to password managers is that they do have a small learning curve that may not be worth it to you. A password book is a great alternative. Many bookstores sell password books, but it could also be as simple as a notebook with a list of sites and passwords. This book should be kept in a secure location and if you use the hymnal method described above then only storing the hymn number will also add a level of security.

The one common practice that you should avoid is a spreadsheet on your computer with all your account credentials. If an attacker ever gets access to your computer they will look for a file with passwords and you will quickly have a much bigger problem on your hands.

What is Hashing?

Hashing is a topic that is going to come up again in this book, so let's take a quick look at how hashing works. If you're not ready for that right now, feel free to skip ahead to the next chapter.

Hashing is a method to create a unique fingerprint (or hash) for any information that you pass into the hashing algorithm. Any time you pass in the same information you will always get the same hash out. We can use this to verify passwords or make sure that files have not been altered. We can even use a hash to identify certain known files (which will come up again in the chapter on antivirus software).

To understand how this works, lets create our own simple hashing algorithm. Picture a chess board that goes on forever in all directions. Here are the rules in our simple hashing algorithm:

- Start in the center of the board
- Move left each time you see a consonant
- Move right each time you see a vowel
- Move forward every time you see a number
- Move backward every time you see a special character

When a user sets their password our website will go through each character in the password and follow our hashing rules. In the end our algorithm will give us the final location of our chess piece and we are going to make note of that final location for the user and then forget the password.

We no longer know how we got to that final location; we only want to keep track of the final location.

The next time the user logs into our website we will take the password they use to login and run it through the same rules. Once we get to a final location we will check that against the final location we stored with the user. If they match, then the user is allowed to login.

With these simple rules I'm sure you can imagine how two different passwords could lead to the same location. The rules used by modern hashing algorithms are significantly more complex and it is incredibly rare to get the same output for two different inputs.

Some websites will take an extra step to protect their users' passwords called "salting". Using our chess board from before, the basic idea of salting is that I am going to decide to start my rules from a different point other than the center of the chess board. If I always start two squares down and one square to the right then I can still use all the same rules for storing and testing passwords, but my hashes will be different than hashes from other sites, even if the password is the same. This protects the passwords that you use on my site from matching the password hashes stored in an attacker's rainbow tables from another site.

3

Multi-Factor Authentication

We'd love to think that every website will treat our account with the best security available, but the reality is that leaks happen. We've talked about the importance of using unique passwords for each site so that a leak on one site doesn't compromise your account on another site, but there's more that you can do to keep your accounts safe.

Multi-Factor Authentication (or MFA) adds layers to your account security. The basic idea is that we can break down your login options into three categories:

- Something you know
- Something you have
- Something you are

Passwords fall into the "something you know" category, along with PIN numbers and the answer to security questions.

For most people, "something you have" is usually your phone. Whether a website is sending a text message with a code that you need to enter or a pop-up message that you need to click on, the result is using something that you routinely have with you as part of the login options.

The "something you are" category is less common for most people, but cell phones have been leading the way with fingerprint and facial recognition. Laptops have also started to add these features and you will probably see more options (often called biometric authentication) in the near future.

When we talk about multi-factor authentication we are talking about making an account more secure by requiring that you use options from two or more categories to login to an account. For example, your bank might send you a text message with a security code to use after you login with your password. Another example might be when your cell phone fills in your saved password for a site after first prompting for a fingerprint.

Basically, as it gets easier for an attacker to get past one authentication factor, you can always make your account more secure by using more authentication factors. Many websites will now give you the ability to set up a second factor for your account, and some are starting to require that you have two factors. In the future you will also see more sites moving away from passwords as one of the factors as they start to favor the other two categories.

Of course, attackers are not sitting around and watching the world change around them. They have come up with a number of ways to take advantage of MFA, so we'll go through the most common ones here.

The Direct Approach

The most common way to get someone's security code is to just ask them for it. You might think that this couldn't happen to you, but attackers are getting very good at this option. They may call pretending to be from your bank and asking to verify your identity before they give you some important information about your account.

In the example shown below, the attacker found a classified ad and found the seller's information in a previous data leak, so they already had the user's e-mail password. However, the account was set up to send a security code to the user when they tried to login, so the attacker posed as a potential buyer who wanted to verify the seller. They informed the seller that they were going to send a code from a different number to verify their identity and then simply logged into the user's e-mail account, triggering a code from the e-mail provider. When the user's e-mail provider sent them the security code, they sent it to the attacker and the attacker was able to login to their e-mail.

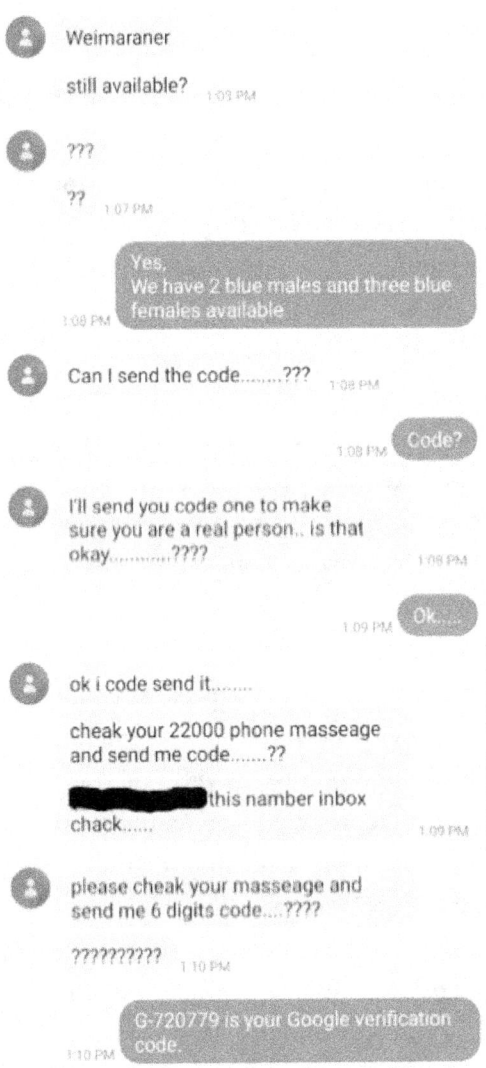

Weimaraner

still available? 1:03 PM

???

?? 1:07 PM

Yes,
We have 2 blue males and three blue
females available
1:08 PM

Can I send the code.........??? 1:08 PM

1:08 PM Code?

I'll send you code one to make
sure you are a real person.. is that
okay............???? 1:08 PM

1:09 PM Ok.....

ok i code send it........

cheak your 22000 phone masseage
and send me code.......??

████████ this namber inbox
chack..... 1:09 PM

please cheak your masseage and
send me 6 digits code....????

?????????? 1:10 PM

G-720779 is your Google verification
code
1:10 PM

This type of attack is becoming very common, especially
if you sell stuff online. Let's talk about how to avoid this.

Security codes are used to verify your identity when you call someone or login to a site – they should **never** be given to someone who called or contacted you. If someone is posing as your bank or a retailer and pressuring you, hang up and call the number from the back of your card or the customer service number on their website to verify that the person calling you is legitimate.

MFA Fatigue

This attack is so simple it's stupid, but it works most of the time. While a text message is still the most common second factor, using an app on your phone that sends a pop-up to approve the login is growing in popularity. Attackers know this and have started to exploit this option as well.

If an attacker already has access to your username and password they can just send login requests over and over, sending hundreds of pop-up notifications to your phone. They are hoping that at some point you will just approve one so the pop-up messages go away. In one recent case they have even called the user posing as a member of the user's IT department and asking them to approve the login because there was a problem with the login software and the pop-ups would keep coming until the user approved the login.

The fix for this is as simple as the problem. If you get login approval pop-ups for an account that you are not trying to log into then just ignore them. If they keep coming then turn off your phone and unplug for a few hours until they give up.

SIM Swapping

First some background – every cell phone has a SIM card inside the phone (it may be a physical card or virtual) which tells your phone company that this is your phone on your account and tied to your number.

When banks started using text messages to send security codes to users, attackers learned that if they could convince your phone company to change the SIM card associated with your account then your calls and text messages would get routed to their phone and they could get access to your bank account. (See the chapter on "Social Engineering" to better understand how they might do this.)

These attacks are difficult to pull off and less common, but there are still a couple of things you can do to protect yourself from this type of attack. First, you can make sure the login to your cell phone account with your phone company is secure so that an attacker can't just log in and make the change on their own. Second, most cell phones have an option to set a SIM PIN number separate from your phone's PIN number that will make it harder for an attacker to switch your number to a new phone. Finally, if you notice your phone suddenly doesn't have any service you should contact your cell phone company right away to understand why it stopped working.

4

Social Engineering

There once was a time when attackers used sophisticated means to gain access to computer networks or crack passwords. A lot of these may still work (though probably not as well as they do in movies), but attackers have found that the easiest way to get information out of people is to simply ask them. This created a category of attacks that we refer to as "Social Engineering".

Another way to think of social engineering is hacking people instead of computers. An attacker could spend a couple of days trying to crack your password, or they could just ask you for it. The attacker may struggle to get a malicious program installed on your computer, or they could just e-mail it to you with installation instructions.

Attackers will often use pressure tactics or other ways to keep you from thinking too much about what they are doing. They may try to make you feel sorry for them. There are some great examples of social engineering in action on YouTube if you search for "What is Your Password? Jimmy Kimmel Live" or "This is how hackers hack you using simple social engineering oracle mind".

In both cases the "attacker" was able to take advantage of the fact that we often want to be helpful and may give away too much information or do something that we might not otherwise do. We are a lot easier to hack than most computer systems, but if you can recognize a few basic tactics then you can recognize when someone is targeting you and save yourself from being taken in.

Phishing, Spear Phishing, and Whaling

Let's start with something that most people have dealt with – a phishing e-mail. Phishing e-mails are usually sent out in bulk by an attacker hoping that they can get a few people to click. They may include your name or other publicly available information to give the e-mail more credibility.

There was a time when I would give you a long lecture on looking for typos in an e-mail or links that don't look right. Chances are that you have already been through that training at work. The problem is that attackers are getting better. They have new tools to hide the links that you click on. They will often work with translation companies to make sure that the e-mail is worded correctly. There are technical things that your e-mail provider is probably doing to keep phishing e-mails out of your account, but we're going to focus on recognizing common tactics instead.

Some e-mails will focus on getting you to log into a

website. The attacker may have set up a site that is an exact copy of a site that you would use every day, like a social media site or your bank site. The safest thing that you can do if you think that an e-mail might not be legitimate is to go to the website instead of clicking on the link. You can also call the company for more information or to verify the e-mail.

Saving passwords in a password manager can also save you in this situation. If you have saved your password for Some-Website.com in your password manager then your password manager will only fill in the password on SomeWebsite.com. If you get a malicious e-mail trying to get you to login to S0meWebs1te.com then your password manager will not fill in the password. This should be a red flag that you are probably on a malicious site.

Some e-mails will include an attached file that tries to install a virus. Antivirus software is much better at stopping these files, but attackers have a couple of tools they use to get around this that can also serve as red flags.

Password-protected files are often used to keep your antivirus software from inspecting a file for malicious content. If you receive an e-mail with an attached file and a message that you will need a password to open the file then that could be a red flag. If the e-mail also contains the password then that will almost always be malicious. However, your doctor or HR department might send you a file that requires some personal information (like a birthday) to unlock. Using a reference to personal information that you both know without including the password itself is a common technique used to securely share password-protected confidential information.

Macros are a tool sometimes used to make documents or spreadsheets more functional. However, attackers learned they could also use macros to put malicious code in a file that would install a virus on a user's computer. Because of this, most computers are set up not to run macros unless the user allows them to run, so attackers will send a message about how you must turn on macros for their really cool spreadsheet to run. Files that require macros should be a red flag that the file is probably malicious.

Some attackers have resorted to old fashioned extortion in their e-mails. The basic format for these extortion messages is that they installed a virus and they are watching you. They captured you doing something you wouldn't want your family to know about, but if you send them money (usually as Bitcoin) then they won't tell anyone. These are most common right after a data leak from a big website because they will use information from the leak that they claim to have learned about you from your computer to make it seem like they are really in your system. They may even make the e-mail look like it came from your e-mail account. They aren't in your system, and they didn't capture you doing anything. Just delete the e-mail and move on.

Using information from data leaks to personalize an e-mail is a technique called "spear phishing". An attacker could also look through your social media profiles or several other online sources for information about you. There is more information about you online than you probably realize, so don't let an attacker get in your head by including personal information that you don't think they could know about you.

One phishing attack that is becoming more common in the workplace is called "whaling". The basic idea is that you get an urgent e-mail from someone in upper management asking you to do something, so you feel pressured to do it now so you don't lose your job. Usually this takes the form of "I am at a client's site and I need you to go buy a stack of gift cards and send me the numbers right now so we don't lose the client." I've never seen a client relationship saved by a stack of gift cards, but if you think there is some reason that the e-mail might be legitimate then it is time to pick up the phone and call to verify the request. Attackers may put a fake signature in the e-mail with their phone number, so always make sure you are using a number you trust. This might include looking up the person's number in the company directory or calling their assistant to verify their travel plans.

Vishing

When an attacker uses phone calls instead of e-mail then we call that "vishing" (for voice phishing). These callers can range from someone pretending to be from your computer company letting you know that they detected a virus on your computer to someone pretending to be from the IRS trying to help you with a tax-lien. These usually fall into two categories.

The first type of call is an attacker who wants to get on your computer. They may pretend to be from tech support calling to help you fix a problem or remove a virus from your computer. Once they get on your computer they will usually lock you out, go through your computer looking for information, and then, once they have everything they want, they will demand money to unlock your computer. They are usually looking for banking information or password files they can use to keep attacking you in other ways.

As someone who shops around for tech support services that will call you when they detect an issue, I can promise you that no software vendor is doing that for you for free. They did not detect a problem on your computer and you should never let anyone connect to your computer remotely for any reason unless it is someone you know and trust.

The second type of call is an attacker who claims they are trying to help you resolve a financial situation. These range from the IRS calling to help you with a tax issue to a police department calling to help you avoid going to jail over an unpaid fine. They will then give you the option to use gift cards to pay the fines. They may even find a way to convince you that they sent you too much money and they just want you to send the difference back.

There are two things to remember when someone tries to exploit you like this: government agencies like paper trails so they conduct business using letters instead of phone calls, and they don't do business in gift cards. Either of these is a red flag that someone is trying to scam you out of money.

These calls use high pressure tactics to keep you from thinking too much about what's happening and they are hard to hang up on. They may even call back several times. Sometimes it takes a good lie to make them go away. Do what you have to do to get them off the phone and block the number.

Smishing

When an attacker uses text messages this is called Smishing (for SMS phishing). These attacks can follow any of the patterns that we have already discussed, but they are most common in cases when people conduct business by text messages, such as when you list an item for sale online or if you run a small business.

One tactic that comes up more with smishing is the idea of overpaying and asking for a portion of the money back. The basic approach is that a buyer wants to buy what you are selling but they are out of town and will need a friend to pick up the item. They want to send you the money plus some extra money for you to give the friend so they can ship the item back. Once you give their friend the money their original payment will disappear (usually a check that doesn't clear or an online payment that gets pulled back) and you are out the money that you paid the friend and the item that you were trying to sell. Any time someone wants to overpay and get money back they are probably trying to scam you.

TOAD Attacks

Telephone Oriented Attack Delivery (TOAD) attacks are another rising trend. These attacks combine a variety of the methods above, but the end result is that the attacker wants you to call them because that gives them more credibility.

These usually take the form of an invoice for an expensive item that they want to confirm before the charges go through or a shipping notice that they want you to call and check on.

There is usually a high dollar amount involved so you feel pressured to call. Once you call they will then ask for personal information like your credit card details or social security number to verify your identity. They are trying to collect this information to use or sell and they are hoping that you will trust them enough to give it to them because you called them.

Regardless of the technique an attacker is using, if you ever feel pressure to do something that doesn't seem right then you should always stop and take a breath and look at the whole situation. Call a friend and ask if the situation seems right. Call your bank and ask them if a situation seems right – they are trained to help you recognize this kind of attack. The attacker is going to try everything they can to keep you from doing this, but hanging up on them and blocking their number is always an option.

5

E-Mail

Most of us look at our Inbox and wonder how it ever got away from us, but to an attacker that's the holy grail of information and access, so let's talk for a little bit about what an attacker sees if they get into your e-mail.

If you're like me you've probably accumulated a bunch of welcome e-mails with a lot of helpful information about your new account. These may include an initial password that you have probably changed by now, but this still let's an attacker know where you have accounts.

Most online accounts also have an option to reset a forgotten password that will e-mail you a link to reset your password. If an attacker can look through your e-mail to find where you have accounts and then use that same access to your e-mail to reset your passwords then they can just change your password without ever knowing what it was. Setting up multi-factor authentication (MFA) makes that process a lot harder most of the time, but that's not always a foolproof backup. Many online accounts will have an option to reset your MFA in case you lost your phone, and that process usually relies on e-mailing you a link. Your e-mail really is the key to your digital life.

We've already talked about passwords and MFA, so there's no need to dive too deep into those areas again. You should have a strong and unique password for your e-mail account (such as your Gmail or Yahoo account), and you should use MFA for the account. The first thing an attacker is going to try when they get their hands on a data leak is to use any information or credentials they find about you to get into your e-mail account.

Of course, there is a good chance that you may find out that your credentials leaked somewhere, or you may get notified about suspicious activity on your e-mail account and you are going to pick a new password that is strong and unique. This time you're even going to remember to set up MFA.

The attacker knows this, so let's talk about what they're going to do the second they get access to your account. If you go looking in your account settings for your e-mail there is probably a section for "Forwarding" and another one for "Filtering" or "Rules". These are different tools that are intended to help you manage your inbox. You may use the forwarding tab on an old account to send all your e-mail to a new account. You may use filtering rules to automatically archive e-mails from a certain e-mail address. Attackers know that most people either never look at these rules or set them and forget them. They will use these settings to hide in plain sight. They may do something as simple as adding their address as

a forwarding address so they get a copy of all the mail you receive. If they are targeting a specific account they might add a
rule that sends them a copy of any email from yourbank.com.

If you ever have any reason to believe that someone has
accessed your e-mail account then you should look for forwarding addresses or filtering rules that you do not recognize.
If you see any rules delete them immediately and change your
e-mail password.

6

Social Media

We like to be social people and social media has given us a way to express that need online. But like anything that we do online, it is not without its risks.

We tend to share a lot on the assumption that we are sharing amongst friends, but the reality is that social media is not as controlled a space as we think it is and once we put information online we lose control over where or how it is used.

Just Another Application

We've talked about the fact that websites have leaks and what those leaks can mean for your passwords and other account information. Social media is just another application and social media sites have leaks too. Remember to use a good password and MFA options for your social media accounts.

Privacy Settings

In your social media profile settings there is usually a section marked "Privacy" that gives you some ability to control who can see the information you share. These settings can change as the application grows and adds features, so you should go back in a couple of times a year and check these settings.

As you do, one of the options is usually "Friends Only" or "Friends of Friends". Just remember that if you have 100 friends and they have 100 friends then "Friends of Friends" works out to be about 10,000 people, 9,900 of which may be strangers.

Oversharing

Part of being social is sharing. We love to let our friends know how we're doing and we don't always think about what we are sharing or how widely we are sharing personal information.

Let me start with a real-life example and then we'll get back to social media. I was sitting on a shuttle bus trying to get to the airport when another passenger near me called their friend. Apparently a friend was picking up the car so they wouldn't have to pay too much for parking while they were out of town. The passenger told the friend what spot the car was parked in and where they hid the car key.

So now I know: the passenger will be out of town, where their car is parked, and where I can get a key for the car. There is a very good chance that if I look in the glove box then I can also get a home address and there might even be a house key on the same keyring.

With that in mind, let's talk about the information that attackers want to know about us. When you create an account most accounts ask you to set up security questions. These questions are used when we reset our passwords or when we call customer service to make changes to our account. They should be treated with the same care that we use with our passwords.

However, just like with our passwords, sometimes we will give the information away if someone just asks us nicely, such as a post on social media.

"What was your first car? Thousands can't recall! Can you? Challenge your friends!"

If you play along you may have just shared a valuable piece of personal information. Who can see it? Was it just on your friend's wall? Odds are good that you saw it on their wall because they answered the question too, but neither of you know anything about the page that initially shared the meme and is now collecting responses.

It's time to stop sharing so much personal information. It's time to stop taking surveys that give unknown pages full access to your profile. It's just not worth it to find out what your spirit animal would be if you lived on the moon.

Profile Duplication

Every week or so one of your friends will post "I've been hacked – don't accept any friend requests from me". It's important to understand what's actually going on here so you know what to do when it happens and how to prevent it.

The profile was probably not hacked, just duplicated. If your name and profile picture are available for an attacker to see, either because of your privacy settings or because you answered a poll that they posted, then an attacker can copy your name and picture and create a new profile with the same name and picture.

They will then look at your friends list if it is visible (based on privacy settings) or look at the people that responded to your answer on their poll so they can send out connection requests to your friends. If your friends reply to the connection request, the attacker may try to get information from them, but usually they just go into some random sales pitch for an amazing investment opportunity. My typical response is to try and throw them off by saying "Thank you! You were never this helpful when you were alive!" which usually makes them go away and often abandon the fake profile altogether.

If you receive one of these requests the best thing to do is report it as a fake profile to the social media company. On the report you will tell them who you think is being impersonated and they will contact that person to verify that the profile is fake and then delete the profile.

Contests

Amazing giveaways have become an ever-present part of social media spam. Enter here for a chance to win a lifetime supply of your favorite chicken nuggets or ice cream!

These are almost always a scam to get you to provide

personal information or access to your social media profile so that an attacker can create a duplicate profile.

The first key to recognizing a scam is knowing that companies put a lot of effort into cultivating their social media presence. They are going to run contests on their own page instead of spending the money to build up a temporary page for the contest.

When you see a contest option there is a link to the page running the contest. Click that link to open up the page and see how many posts they have. If the only post is the contest then it is a scam. If there are hundreds of posts going back years then it is probably the company's page and a legitimate contest.

7

Wireless Networks

Everyone these days seems to always be on the lookout for Wi-Fi. We rely on wireless connections to keep our social media scrolling and our movies streaming. Our devices try to help us out by remembering networks that we have used in the past and automatically connecting when they see that network again.

There are two problems with that: our devices aren't very good at distinguishing between networks with the same name, and attackers know how to take advantage of that fact to get our data.

Wireless Spoofing

One of the companies I worked for would insist that every employee use the same hotel chain when we traveled. That hotel chain used the same wireless network name in all their hotels. Anyone who had ever traveled for work had connected to that network and that connection was now saved on their device.

One of the things I liked to do when I needed to remind people about wireless security was to set up a wireless network in the office cafeteria that used the same network name as the hotel chain. Within minutes devices from all over the company would start to connect because our devices rely on the network name to find networks they know.

Attackers know the network names for a lot of public networks like coffee shops, hotels, and airports. They will use this to sit out in public places and set up rogue networks using these common names and wait for users to connect. Most users are happy to have Wi-Fi and won't even check to see what network they are using.

Once a user connects to a rogue network the attacker can see all the traffic that goes through the network. Fortunately, most websites encrypt that data and the attacker can't read it, but there is always some data that doesn't get encrypted and sometimes that can be useful to an attacker.

An attacker can also try to redirect the user's traffic to a fake version of a website to trick a user into logging in and giving the attacker their login information. Your browser will usually detect this and give you a security warning, but there are always people that will click through the warning to visit the site.

When you finish using a public network and before you leave the area, it is always a good idea to go into your device's Wi-Fi settings and select the name for the public network. There will be an option in the settings to forget the connection. This will prevent your device from connecting to that network, or a rogue network using the same name, in the future.

Virtual Private Networks

Virtual private networks (or VPN) are another tool you can use to secure the wireless traffic on your device. If you think of the internet as a giant pipe passing data back and forth then a VPN acts as a smaller pipe that you set up between your device and your VPN provider to protect your data from everything else in the pipe.

This smaller pipe is always encrypted, so if an attacker is watching the traffic on the network then they will not be able to read your traffic. This provides an inexpensive way to make sure your data is protected even if the network you are using is not.

With most VPN providers you will set up an account for a small monthly fee and they will give you an app you can download to your personal devices. When you run this app and login to the VPN the app will build this secure tunnel between your device and the VPN provider. The VPN provider will still see your traffic, but nobody in between you and the VPN provider will be able to see your traffic.

I've heard a lot of VPN commercials lately so I wanted to also take a second to let you know what a VPN is not. VPN hides your traffic in the pipes, but if you connect to a website, then that website will still track your session. They will not know who you are unless you login or provide them with that information, and they will not know your home IP address or where you are connecting from, but you should not assume you have complete anonymity just because you are using a VPN.

8

Personal Finances

There was a time that I used to worry constantly about whether or not my credit card would be compromised. I've learned that it is easier to make plans to identify when my card is compromised and how I can best respond, because at some point we will all likely have to deal with this situation.

Skimmers

Skimmers have become a reliable tool for stealing credit card information. An attacker places a device inside a credit card reader that reads the information from your card's magnetic strip as you swipe it in the card reader. They are getting smaller and harder to detect and they are more common in card readers that are available to the public and not closely monitored, like gas stations.

One of the defenses against this type of attack is to use contactless cards or contactless payment options on your phone.

These are harder for most skimmers to read and may also use a temporary code instead of your card's actual information.

Another option is to use a gas card for gas purchases. This does not prevent your card from being skimmed, but it will limit the ability of an attacker to use or sell your information because a gas card has limited usefulness so it is not as valuable to an attacker.

Online Shopping

Shopping online is just so convenient that it has become a part of our daily lives. We hand over our card details on a regular basis and trust that the sites will handle our information securely.

While I try to shop on sites I think I can trust, you've probably guessed that I don't like to rely entirely on trust. If a site accepts a third-party payment processor like PayPal, Google Pay, or Apple Pay then you can reduce the number of sites that have your payment information, thereby reducing your exposure to potential data leaks.

Some banks or third-party services like Privacy.com offer temporary virtual credit cards which can be used once for an online transaction and then the card immediately becomes invalid. This protects you from potential data leaks because the information will not be usable if the site's information is leaked by an attacker.

Transaction Monitoring

Many banks now offer apps that will give you an alert each time your card is used. This can give you a heads up if your card information has leaked online. Let's walk through what an attacker might do once they have your information by looking at the last time my card information leaked online.

The first thing an attacker wants to know is if the credit card information they either stole or bought from another attacker is valid. They will want to do this without being detected, so they will typically start an online transaction for less than $20 and then cancel the transaction before the card is actually charged. On my bank's app I saw two transactions pop up that I did not recognize. They were both under $20 and both pending (which means that they never settled). I called my bank immediately and they cancelled my cards and began the process of sending me new cards.

In just under two hours I began to see fraudulent transactions hit my card. Fortunately the card had already been shut down and the transactions were declined. Had I not recognized the two test transactions and contacted my bank I would have been on the hook for almost a thousand dollars in fraudulent transactions. It is not enough to review your bank statements at the end of the month, you need to keep an eye on them in real time.

Credit Cards vs. Debit Cards

Your bank has policies that determine how they are required to respond if your card information is stolen. One thing they do not highlight is that the policies are different for credit cards and debit cards.

If you report a fraudulent transaction on your credit card your bank will typically return the funds to your account while they investigate the fraudulent activity. The issuing bank is responsible for any money that was stolen.

If you report a fraudulent transaction on your debit card then you are responsible for the money that was stolen. The bank will still do everything in their power to investigate the transaction and return the funds, but that money is gone unless the bank is able to confirm that the transaction was fraudulent.

Your bank's terms may vary, but it is important to understand that not all transactions are treated the same. You may want to consider the type of card that you are using when dealing with riskier transactions like small sites with questionable security.

9

Odds and Ends

There are a number of random items that didn't really fit into any chapter, so I'm going to collect them here.

Security Questions

We talked about security questions back in the chapter on social media, but there is still more we can do to protect our personal information. When you are setting up an account and enter the name of the city where you were born, nobody pulls your birth certificate to check. It's ok to pick a city that doesn't match your birth certificate – as long as you make a note of your answers for the next time you need them.

QR Codes

QR Codes (or Quick Response Codes) are those square bar codes that you can scan with the camera on your phone and they are popping up everywhere. They are primarily used for sharing links, but they can contain several different types of information. They are an incredibly convenient way to get information onto your phone and we often scan them without even giving it second thought.

Attackers have taken note of how easily we scan any QR code that we see and they will use this tendency to get us to open malicious links. Attackers have started creating stickers that will cover legitimate QR codes with QR codes that point to malicious links. Before you scan a QR code, stop to see if it is a sticker. Many phones will also show you the link before opening your phone's browser, giving you the chance to make sure you are opening the website you expect to open with the code.

Antivirus

Most Windows computers now come standard with antivirus, and many internet providers now also include it as part of the base package or for a small additional fee. If not, you can purchase an antivirus subscription for a reasonable monthly or yearly rate.

Most antivirus software works using the same hashing method we discussed back in the passwords chapter. The antivirus software has a database of file fingerprints for known viruses. When it finds a file that matches the file fingerprint for a known malicious file it will delete or quarantine the file.

As you might imagine, new malicious files are identified

and added every day, so it is important to make sure that your antivirus software is updated on a regular basis.

Rebooting

Your computer is also checking and downloading updates in the background on a regular basis, but it cannot always install these updates while your computer is up and running. It is important to remember to reboot your computer about once a week so that it can install any important security updates.

10

What to Do if You Are in a Data Breach

Data breaches are an unavoidable part of life these days, so we're going to talk about what to do the next time you get an e-mail from a website letting you know that they've lost your data. Even though website owners are typically required to notify you if they have leaked your data, you should still use a breach monitoring service like the free Have I Been Pwned website (haveibeenpwned.com).

Stay Calm

Data breaches are becoming common these days. As of December 2022, the Have I Been Pwned website reported roughly 12 billion accounts as part of their database of breached accounts. In other words, you're in good company.

Learn About the Breach

The company will often send you information about the specific data elements in the breach, but you can also typically find the breach on a website like haveibeenpwned.com and get additional information about what was leaked.

The correct response for any data breach is specific to what type of data has been stolen. Here are some common data elements and suggested responses:

- E-Mail, Username, Full Name, other biographic information: You may start receiving new junk e-mail, but for the most part this data is pretty common knowledge anyway and you should assume that it is out on the web. Keep that in mind when choosing security questions for websites and change any that may rely on this information.

- Plain Text Passwords: Change your password for this site and any other site that uses the same password. Hackers will often add your account to a database and use it to try other websites (social media, banking, etc.) to see what else they can get into.

- Hashed or encrypted passwords: It takes more work to decrypt an encrypted or hashed password, but computers are getting really fast these days. It may take some time for them to get your password, but you should go ahead and assume that they can and follow the same guidelines as plain text passwords.

- Credit Card, Banking, Social Security, or Other Financial Information: Contact your bank or financial institution immediately to replace credit cards, check for unrecognized transactions, and any other security protocols they have in place. The sooner you contact your financial institution the sooner it becomes their problem. Sign up for credit monitoring if you don't have it already and check your credit report for accounts you don't recognize.

General Tips

Here are some general tips and tricks to consider (we've already discussed some of these, but this puts everything in one place):

- You can reduce your risk by using unique passwords for every site. The best password is the one that you don't know. Consider using a password manager like LastPass or 1Password so that you can use a password for each site that is unique and random.
- After each data breach there is an uptick in blackmail e-mails that use the information in the breach to convince you that they have access to your computer. They don't, so don't pay them.

There was a time when the focus was on preventing your data from being breached. While this is still a good practice, we've reached a point where a lot of people collect data about you and not all of them will be as careful as you are with that data. Take some time to think about how you can minimize the impact when your data might be leaked because there is a very good chance that data about you has been leaked or will be leaked soon.

11

Conclusion and Checklist

Online security will constantly be changing and nobody can tell you every technique that attackers will use. However, I hope this book provides you with enough insights into how an attacker works that you can avoid looking like an easy target or recognize when you are being targeted in time to react.

The following checklist is a summary of the tips that we discussed. Take a few moments to review it every couple of months to make sure you are still taking those steps necessary to keep your head on a swivel on your digital journey.

Passwords

- Passwords should be longer than just a word
- Passwords should be unique and not shared across sites
- Passwords should be managed using a password manager or stored offline

- Passwords should not be in a text file or spreadsheet stored on your computer

Multi Factor Authentication (MFA)

- Use MFA for accounts when available
- Do not respond to an MFA prompt that you were not expecting
- Do not give your MFA code to anyone who contacts you

Social Engineering

- If you receive a request that you were not expecting, use other safe methods to verify the request
- Government agencies do not work via phone calls
- No legitimate business or agency does business using third-party gift cards
- Verify that you are on the site that you think you are on before entering personal information
- Password protected files sent through e-mail along with the password should be treated as suspicious
- Do not enable macros on a document or spreadsheet
- Look up a company's number from a legitimate source before you call them back

E-Mail

- Protect your e-mail like it is the key to all of your other accounts

- Check your e-mail accounts for forwarding or filtering rules that you do not recognize

Social Media

- Regularly review the privacy settings on your social media profile
- Do not share anything on social media that you would not share publicly
- Do not share information that you are using as an answer to a security question
- Do not respond to contests without verifying the source promoting the contest

Wireless Networks

- Don't connect to networks you do not recognize or that seem out of place
- Forget network settings when you disconnect from a public network

Personal Finances

- Use options other than swiping your card when available
- Use intermediate processors or virtual cards with online sites when possible
- Setup and use your bank's real-time transaction alerts
- Review the terms for your credit cards vs. debit cards

Odds and Ends

- Review the answers to your security questions and update those that are too easily guessed
- Check QR codes for sticker overlays
- Review the link before opening a link from a QR code
- Update your antivirus software regularly
- Reboot your computer regularly